Amazing Math:
Introduction to Platonic Solids

Copyright & Other Notices

Updates, news & related resources from the author can be found at
http://www.suniltanna.com/platonic

Information about other math books by the same author can be found at
http://www.suniltanna.com/math

Introduction

For some time now, I have been tutoring both adults and children in math and science. As a result, I have discovered that many people of all ages have a latent interest (and talent) in mathematics that is somehow never got fully awoken while in school.

Because of my experience tutoring, I have been gradually developing a series of books on mathematical and science topics (a full list of my books can be found on my website at http://www.suniltanna.com). Most of my books are intended to teach specific topics and techniques, but I have also written others intended to awaken a student's interest in these subjects and broaden their horizons.

This book is about a fascinating mathematical topic – Platonic solids – which are a particular kind of 3-dimensional shapes. These shapes are not only important mathematically, but are also found in the natural world, and also have applications in science and technology. However, in my experience most people have never really considered these objects in detail, or as a unified group of related shapes.

When you review the Platonic solids, I am confident that you will recognize some of them – and not just the cube – for example, if you have ever played a role-playing game such as Dungeons & Dragons, the shapes of most of the polyhedral dice are all based on Platonic Solids.

In this book, among other things you will learn:

- What the Platonic solids are

- The history of the discovery of Platonic solids

- The common features of all Platonic solids

- The geometrical details of each Platonic solid

- Examples of where each type of Platonic solid occurs in nature

- How we know there are only five types of Platonic solid (geometric proof)

- A topological proof that there are only five types of Platonic solid

- What are dual polyhedra

- What is the dual polyhedron for each of the Platonic solids

- The relationships between each Platonic solid and its dual polyhedron

- How to calculate angles in Platonic solids using trigonometric formulae

- The relationship between spheres and Platonic solids

- How to calculate the surface area of a Platonic solid

- How to calculate the volume of a Platonic solid

- An introduction to some other interesting types of polyhedra – prisms, antiprisms, Kepler-Poinsot polyhedra, Archimedean solids, Catalan solids, Johnson solids, and deltahedra.

Although this is a book about a mathematical topic, please do **not** get the impression that it is aimed only at math experts.   Even if you have only ever studied (or are currently studying) high school math, then you should be able to make sense of this book. Moreover, even if your high school math is a bit rusty, I have even included a brief recap of some necessarily underlying mathematical concepts such as trigonometric operations (sine, cosine, etc.) and radians.

I hope you enjoy this book, which is one of several books that I have written.

- If you want to find out about my math books, please visit: http://www.suniltanna.com/math

- For my science books, please visit: http://www.suniltanna.com/science

Introducing Platonic Solids

Platonic solids are 3-dimensional shapes whose faces are all the same size and shape. There are 5 Platonic solids:

- Regular tetrahedron (also known as "triangular pyramid") - A shape with 4 faces, each face being a triangle

- Regular octahedron (also known as "square bipyramid") – A shape with 8 faces, each face being a triangle

- Regular icosahedron – A shape with 20 faces, each face being a triangle

- Regular hexahedron (also known as "cube") – A shape with 6 faces, each face being a square

- Regular dodecahedron – A shape with 12 faces, each face being a pentagon

In the next chapter, we will develop a more rigorous mathematical definition of Platonic solids, and see what links these particular 5 shapes – and no others.

Let's get started!

Sculpture based of the five Platonic solids in the Steinfurter Bagno park, near Burgsteinfurt, Germany:

What are Platonic Solids?

In the last chapter, I listed the 5 Platonic solids, without really giving too many precise details or explaining why they are related. In this chapter, we will briefly recap some fundamental concepts in geometry, and then see how Platonic solids are defined in terms of these concepts.

You will no doubt recall that a **polygon** is a 2-dimensional shape ("**plane figure**") made by joining straight line segments so as to form a closed loop. A **regular polygon** is one where all the line segments are of equal length, and all the angles between line segments where they meet at each **vertex** (plural: **vertices**) are equal.

Examples of regular polygons include

- **Equilateral triangles** – a 3-sided shape (triangle) with 3 equal sides, and a 60° angle at every vertex

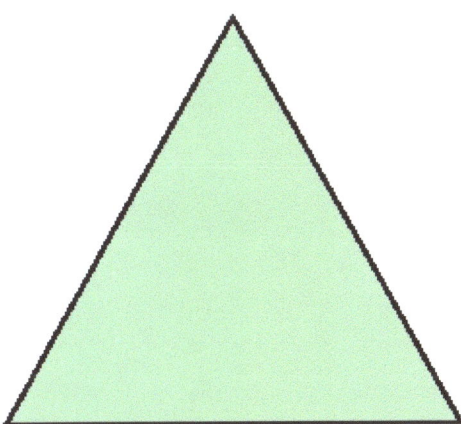

- **Squares** – a 4-sided shape (quadrilateral) with 4 equal sides, and a 90° angle at every vertex

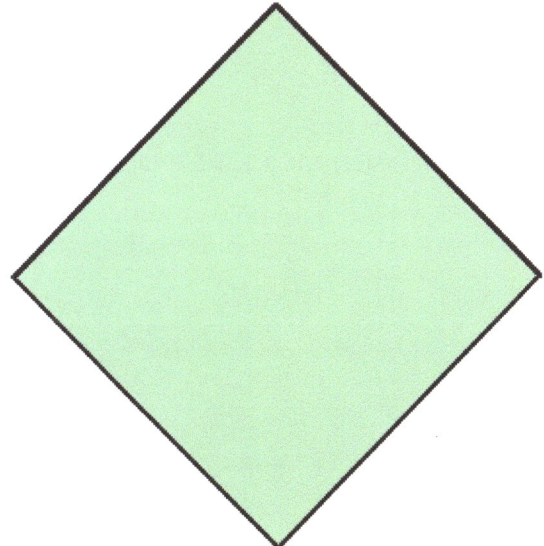

- Regular **pentagons** – a 5-sided shape with 5 equal sides, and a 108° angle at every vertex

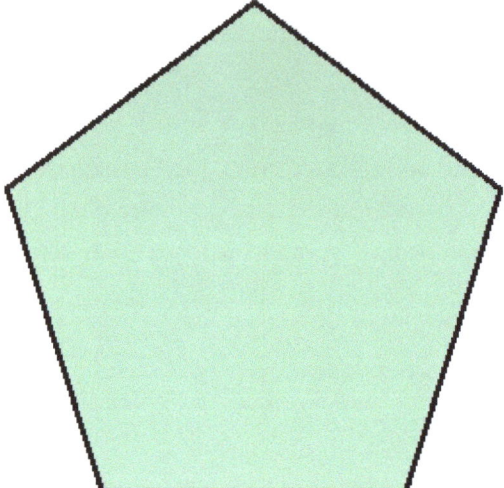

- Regular **hexagons** – a 6-sided shape with 6 equal sides, and a 120° angle at every vertex

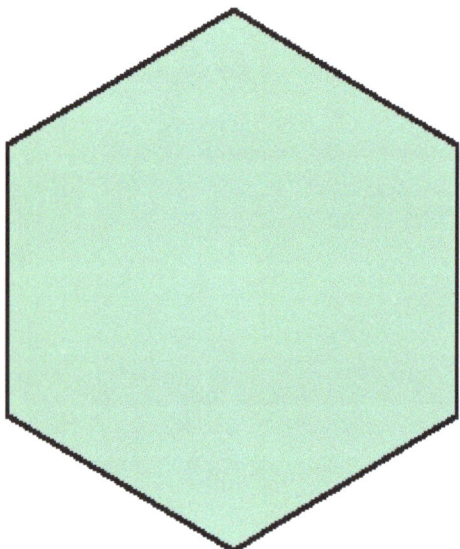

- Regular **heptagons** – a 7-sided shape with 7 equal sides, and a 128.57° (approximately) angle at every vertex

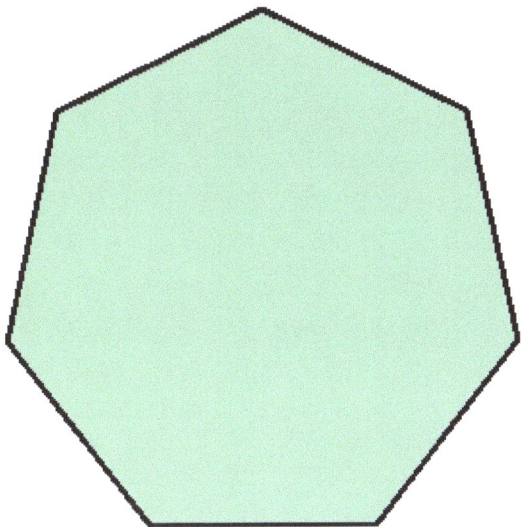

- Regular **octagons** – an 8-sided shape with 8 equal sides, and a 135° angle at every vertex

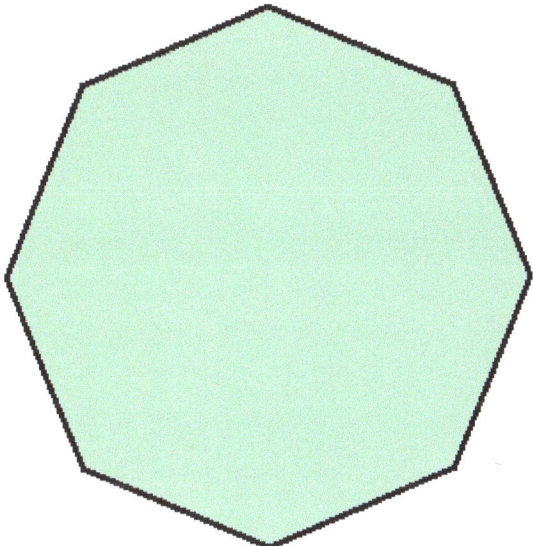

- ...and so on

A **polyhedron** (plural: **polyhedra** or **polyhedrons**) is a 3-dimensional shape made from joining polygons together, with the polygons serving as the **faces** of the polyhedron.

Some examples of polyhedra include:

- **Cubes** (sometimes known to mathematicians as **regular hexahedra**) – a shape with 6 square faces

- Square Pyramids - a shape with 4 triangular faces, and 1 rectangular face

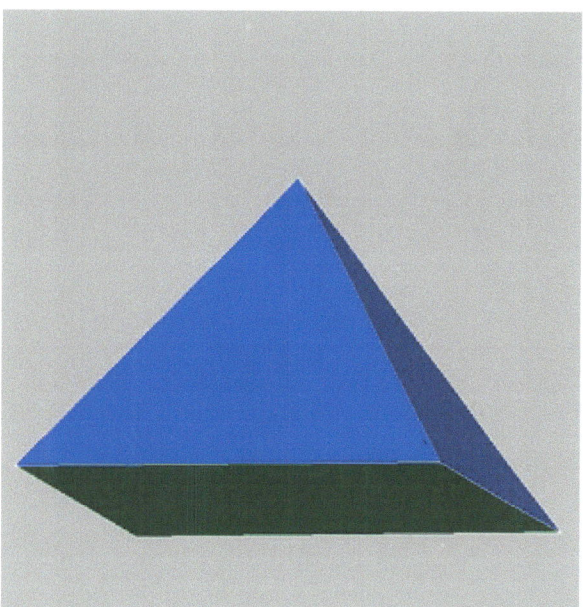

- Triangular prisms – a shape with 2 triangular faces, and 3 parallelogram or rectangular faces

- Hexagonal prisms – a shape with 2 hexagonal faces, and 6 parallelogram or rectangular faces

- Hexagonal antiprisms – a shape with 2 hexagonal faces and 12 triangular faces

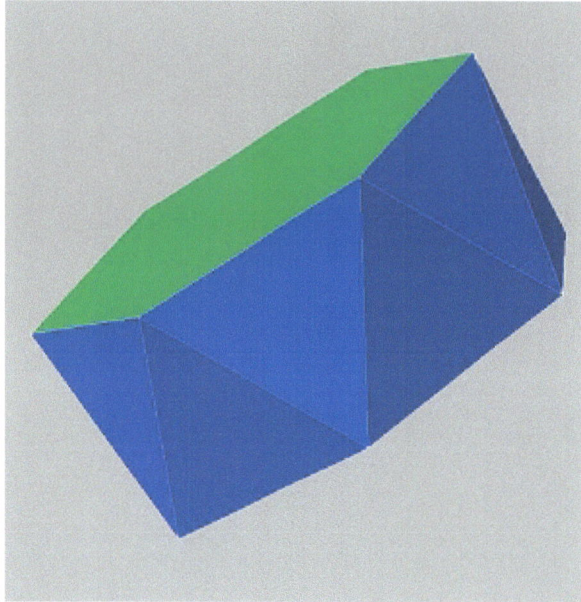

- ...and so on – there are many other types of polyhedra!

A **regular polyhedron** is one which is made of **congruent** (all the same size/shape, although mirror images are allowed) regular polygons assembled in the same way around each **vertex** (corner).

Finally, a **convex** polyhedron is one in which any two points inside the shape can be joined by a straight line segment that itself does **not** go outside the shape. So, for example:

- A cube would be convex, because any two points inside the cube can be linked by a straight line which does **not** emerge from the cube.

- A square pyramid would be convex,, because any two points inside the pyramid can be linked by a straight line which does **not** emerge from the pyramid.

- A triangular prism would be convex,, because any two points inside the prism can be linked by a straight line which does **not** emerge from the prism.

- Any shape that had a hollow or dent in it would **not** be convex (such shapes would be **concave**).   This is because if you choose two points on opposite sides of the hollow/dent, a straight line between these points will be partly outside the solid.

Platonic solids are **convex regular polyhedra**.

What this means in detail is that a Platonic solid is polyhedron with the following properties:

- All its faces are regular polygons – that is to say that the length of all the edges on each face are the same as each other, and likewise, the angles at each vertex on each face are all the same as each other.

- All its faces are the same type of, and same size of, regular polygon.

- Every vertex of the polyhedron is put together in the same way.

- The object as a whole is convex – it does **not** have any dents, dimples, or holes.

Some polyhedra of that fit this definition seem to have been known since prehistory – for example, dice appeared at the dawn of civilization. However, the discovery that there are five and **only** five types of these polyhedra seems to have been first made by the ancient Greeks.

Some people credit the first listing of all five types of these polyhedra to Pythagoras (c. 570 BC to 495 BC), whereas others believe that two of the types of polyhedra were discovered by Theaetetus of Athens (c. 417 BC to 369 BC), who may also have been responsible for the first proof that there can only be five types.

Bust of Pythagoras in the Vatican Museum:

Regardless of who originally discovered all five types, this group of polyhedra eventually become associated with the Greek philosopher Plato (428/427 BC or 424/423 BC to 348/347 BC). Plato linked four of the types with the four classical elements (earth, wind, air, fire), and the fifth with the heavens.   Even though Plato was **<u>not</u>** their discoverer, today the five types of regular convex polyhedra are nowadays collectively referred to as Platonic solids.

Plato:

Next we will review each of the five types of Platonic solid in turn.

Models of the five Platonic solids in the garden of the Weizmann Institute of Science, Israel:

Regular Tetrahedron

A regular tetrahedron (plural: tetrahedra or tetrahedrons) which is also sometimes known as a "triangular-based pyramid", is a polyhedron with 4 faces, each face being an equilateral triangle.

Here is an image of a regular tetrahedron:

Plato associated the regular tetrahedron with the classical element of fire. He did this on the basis that the heat from fire feels sharp and stabbing, which he imagined came from the impacts of the pointed vertices of little tetrahedra.

The geometrical details of a regular tetrahedron are:

- A regular tetrahedron has 4 faces.

- Each face in a regular tetrahedron has 3 edges – so is a 3-sided regular polygon, namely an equilateral triangle.

- There are 4 vertices in a regular tetrahedron, each vertex being formed where 3 faces meet.

- There are 6 edges (formed whenever only 2 faces meet) in a regular tetrahedron.

- The face angle (the angle at each vertex on each polygonal face) is 60°.

- The dihedral angle (the angle between two faces at an edge) is 70.53° (approximately).

- The vertex angle (the angle between edges at a vertex) is 60°.

Here is a **net** (unfolded version) of a tetrahedron:

Regular Tetrahedra and Distorted Tetrahedra in Nature

In chemistry, many molecules have tetrahedral symmetry. This occurs when there is a central atom **covalently bonded** (a bond created by sharing a pair of electrons between atoms) to four other atoms. A tetrahedral symmetry arises because the bonds repel each other, causing them to be evenly spaced, but as far apart in space as possible. The central atom can thus be considered to be at the center of a tetrahedron, and the four surrounding atoms at the four vertices of the tetrahedron.

Model of a tetrahedral molecule:

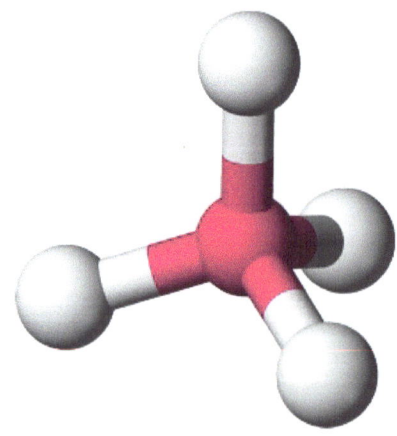

There are numerous examples of molecules with this shape. Some examples of such molecules include methane (CH_4), xenon tetroxide (XeO_4) ammonium ions (NH_4^+), sulfate ions (SO_4^{2-}), phosphate ions (PO_4^{3-}) and perchlorate ions (ClO_4^-), although of course there are many others.

In a molecule such as methane, the four atoms (in the case of methane, hydrogen atoms) around the central atom (in the case of methane, a carbon atom), are all equivalent, and are all bonded in the same way, so the structure is a perfect tetrahedron, and the angle between any pair of two bonds (the **bond angle**) is 109.5°.

There also molecules which can be considered to be a **distorted tetrahedron**, for example, ammonia (NH_3). In the case of ammonia, there is a central nitrogen atom, surrounded by 3 hydrogen atoms (each connected with a covalent bond), and also an additional **lone pair** of electrons (a pair of electrons which are not used for bonding). You can still imagine ammonia as a tetrahedron, provided you remember that one vertex of the tetrahedron is occupied by a lone pair of electrons rather than a hydrogen atom. Additionally, the tetrahedron is distorted in ammonia – because the lone pair is slightly more repulsive than the covalent bonds, the covalent bonds are pushed slightly together, and slightly further away from the lone pair, resulting in a reduced bond angle (as compared to a perfect tetrahedral molecule) of 107.8°.

Model of ammonia molecule (lone pair **not** shown):

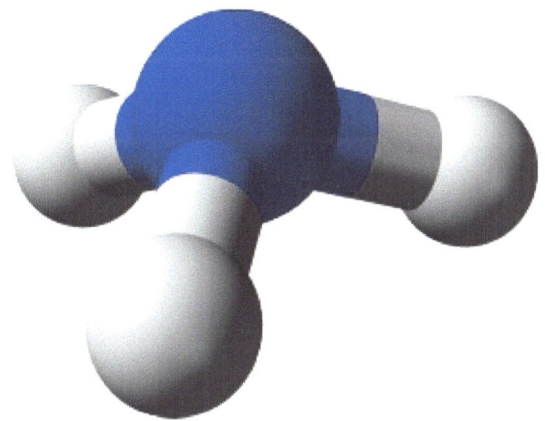

Another molecule whose shape results from a distorted tetrahedron is water (H_2O). In the case of water, each molecule consists of a central oxygen atom, bonded to two hydrogen atoms, but with two lone pairs attached to the central atom.   As there are two lone pairs in water, as opposed to the single lone pair in ammonia, the distortion is even greater, and the covalent bonds are pushed even closer together, reducing the bond angle to just 104.5°.

Model of water molecule (lone pairs **not** shown):

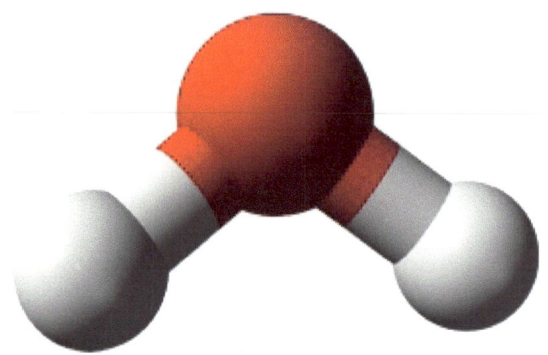

Regular Octahedron

A regular octahedron (plural: octahedra or octahedrons) which is also sometimes known as a "square bipyramid", is a polyhedron with 8 faces, each face being an equilateral triangle.

Here is an image of a regular octahedron:

Plato associated the regular octahedron with the classical element of air. He did this on the basis that air is so smooth that you can barely feel it, and described the octahedron in similar terms.

The geometrical details of a regular octahedron are:

- A regular octahedron has 8 faces.

- Each face in a regular octahedron has 3 edges – so is a 3-sided regular polygon, namely an equilateral triangle.

- There are 6 vertices in a regular octahedron, each vertex being formed where 4 faces meet.

- There are 12 edges (formed whenever only 2 faces meet) in a regular octahedron.

- The face angle (the angle at each vertex on each polygonal face) is 60°.

- The dihedral angle (the angle between two faces at an edge) is 109.47° (approximately).

- The vertex angle (the angle between edges at a vertex) is 60° for edges which are part of the same face, and 90° for edges which are not.

Here is a net (unfolded version) of an octahedron:

Regular Octahedra in Nature

Like tetrahedral structures, octahedral structures also occur in chemistry. Just as a central atom surrounded by four atoms/groups can form a tetrahedral structure (albeit sometimes distorted if the atoms/groups are not all the same), a central atom surrounded by six atoms/groups will form an octahedral structure.   As with the tetrahedral structure, this happens because the surrounding atoms/groups mutually repel, and hence are evenly spaced as far apart as possible, at the octahedron's vertices.

Of course, an octahedral structure can only occur if the central atom can form a stable compound with six surrounding atoms/groups.   The most common situation where this happens is when there is a central metal ion (charged atom) surrounded by six ligands (surrounding ions/molecules), although there are also some octahedral molecules such as sulfur hexafluoride (SF_6).

Model of an octahedral molecule:

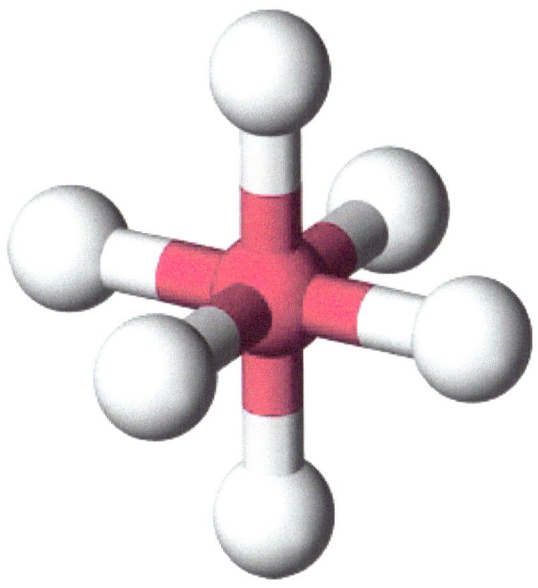

Octahedral shapes and structures also appear in many natural materials. For example, natural crystals of alum, diamond, and fluorite are all commonly octahedral in shape.

Approximately octahedral-shaped rough diamond crystal:

Regular Icosahedron

A regular icosahedron (plural: icosahedra or icosahedrons) is a polyhedron with 20 faces, each face being an equilateral triangle.

Here is an image of a regular icosahedron:

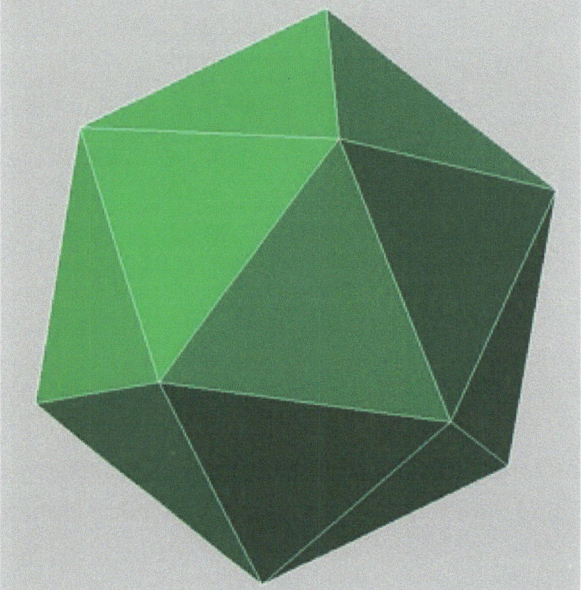

Plato associated the regular icosahedron with the classical element of water. He argued that water flows freely and therefore it must be made of objects resembling little spheres (balls), and he then noted that the shape of an icosahedron is almost spherical.

The geometrical details of a regular icosahedron are:

- A regular icosahedron has 20 faces.

- Each face in a regular icosahedron has 3 edges – so is a 3-sided regular polygon, namely an equilateral triangle.

- There are 12 vertices in a regular icosahedron, each vertex being formed where 5 faces meet.

- There are 30 edges (formed whenever only 2 faces meet) in a regular icosahedron.

- The face angle (the angle at each vertex on each polygonal face) is 60°.

- The dihedral angle (the angle between two faces at an edge) is 138.19° (approximately).

- The **vertex angle** (the angle between edges at a vertex) is 60° for edges which are part of the same face, and 108° for edges which are not.

Here is a **net** (unfolded version) of an icosahedron:

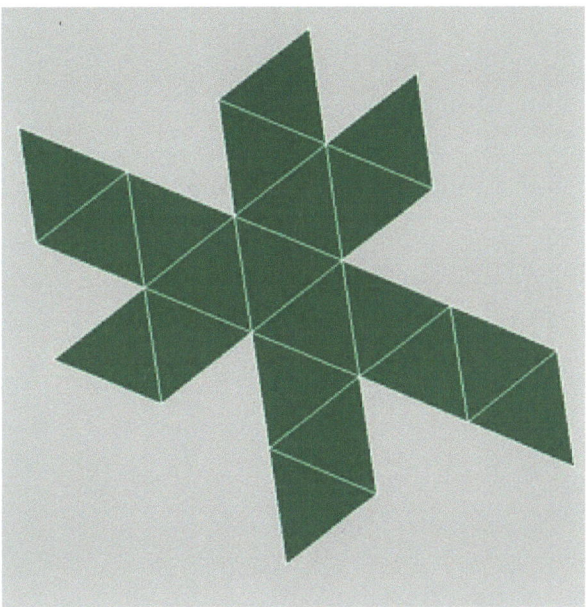

Regular Icosahedra in Nature

Icosahedral structures occur in a number of chemical compounds including closo-carboranes (carboranes are clusters of carbon, boron and hydrogen atoms – formed in the shape of a polyhedron. Some carboranes may be missing one or more vertices of the polyhedral shape, but closo-carboranes have a complete set of vertices), some **allotropes** of boron (solid boron can come in several different forms, each form is known as an allotrope), and many borides (chemical complexes formed from metallic elements and boron).

Icosahedral structures also appear in biology. For example, many viruses (including the herpes virus) have icosahedral shells, bacterial organelles can have an icosahedral shape, and some single-celled organisms including some species of radiolaria (a type of single-celled organism that lives in the oceans) have a skeleton shaped like a regular icosahedron.

Circogonia icosahedra, a species of radiolaria:

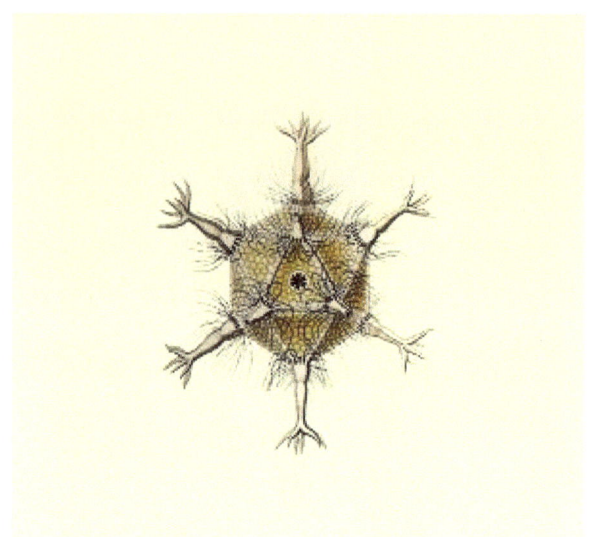

Regular Hexahedron

A regular hexahedron (plural: hexahedra or hexahedrons) which is more commonly known as a "cube" (even among mathematicians), is a polyhedron with 6 faces, each face being a square.

Here is an image of a regular hexahedron:

Plato associated the regular hexahedron with the classical element of earth. His logic was that the shape caused earth to crumble and fall apart when picked up.

The geometrical details of a regular hexahedron are:

- A regular hexahedron has 6 faces.

- Each face in a regular hexahedron has 4 edges – so is a 4-sided regular polygon, namely a square.

- There are 8 vertices in a regular hexahedron, each vertex being formed where 3 faces meet.

- There are 12 edges (formed whenever only 2 faces meet) in a regular hexahedron.

- The face angle (the angle at each vertex on each polygonal face) is 90°.

- The dihedral angle (the angle between two faces at an edge) is 90°.

- The vertex angle (the angle between edges at a vertex) is also 90°.

Here is a **net** (unfolded version) of a regular hexahedron:

Regular Hexahedra in Nature

Crystals and minerals containing regular hexahedra (cubic) shapes are extremely common. This is because they contain repeating units ("**unit cells**") based on arranging atoms in a cubic arrangement.

Cubic pyrite crystals:

There are in fact several different cubic arrangements (involving cubic-type unit cells) in which the atoms within crystals and minerals can be arranged, with different minerals adopting different arrangements depending on their composition and the relative sizes of the atoms involved.

The three main varieties of cubic packing are:

- **Primitive cubic** (also known as **simple cubic**)

- **Body-centered cubic**

- **Face-centered cubic** (also known as **cubic close-packed**)

Solid sodium chloride (NaCl) ionic crystal structure – chloride ions (Cl^-) are shown in green, sodium ions (Na^+) in blue. The structure can be regarded as a two interlocking face-centered cubic lattices (with each ion forming its own lattice), or as a face-centered lattice of chloride ions with sodium ions in the octahedral holes:

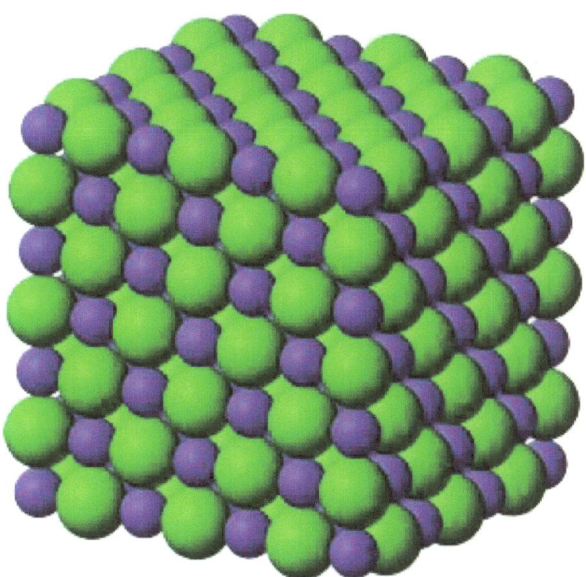

Dodecahedron

A **regular dodecahedron** (plural: **dodecahedra** or **dodecahedrons**) is a polyhedron with 12 faces, each face being a regular pentagon.

Here is an image of a regular dodecahedron:

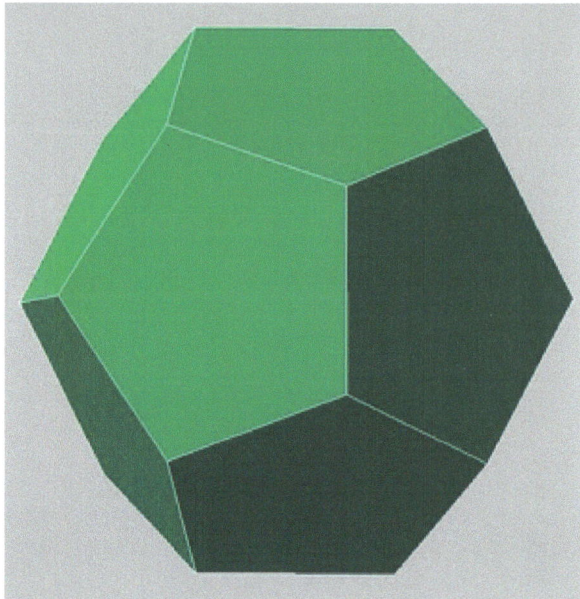

Plato did **not** associate the regular dodecahedron with any particular classical element, but vaguely associated it with the heavens.

The geometrical details of a regular dodecahedron are:

- A regular dodecahedron has 12 faces.

- Each face in a regular dodecahedron has 5 edges – so is a 5-sided regular polygon, namely a regular pentagon.

- There are 20 vertices in a regular dodecahedron, each vertex being formed where 3 faces meet.

- There are 30 edges (formed whenever only 2 faces meet) in a regular dodecahedron.

- The **face angle** (the angle at each vertex on each polygonal face) is 108°.

- The **dihedral angle** (the angle between two faces at an edge) is 116.57° (approximately).

- The **vertex angle** (the angle between edges at a vertex) is 108°.

Here is a **net** (unfolded version) of a dodecahedron:

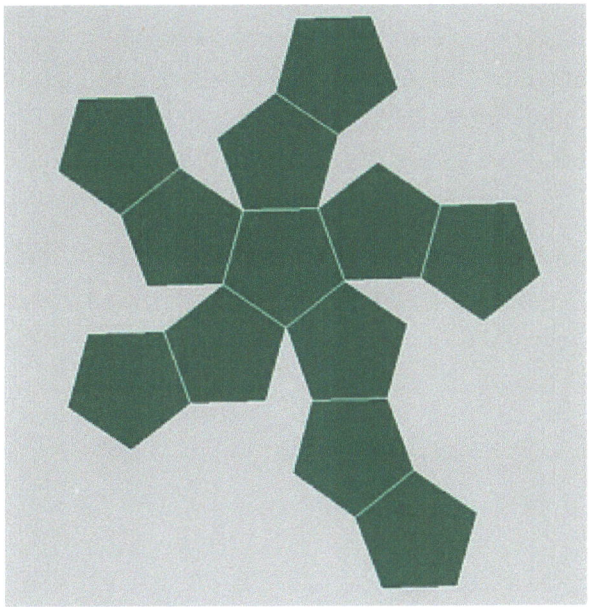

Dodecahedra in Nature

Compared to the other Platonic solids, the dodecahedral shape occurs relatively infrequently in nature. Perhaps the best-known examples of natural dodecahedra are those that occur in some quasicrystals such as Holmium-Magnesium-Zinc quasicrystals.

Holmium-Magnesium-Zinc quasi crystals:

(Please note: when people say that a diamond or garnet exhibits "dodecahedral habit", they are **not** referring to the Platonic dodecahedron – but rather to an entirely different polyhedron, the rhombic dodecahedron. Rhombic dodecahedra are irregular polyhedra which have rhombuses as faces and 14 vertices of two different types. The rhombic dodecahedron is actually one of the **Catalan solids** – a group of polyhedra which will briefly discuss later in this book).

In 2003, French astrophysicist Jean-Pierre Luminet also suggested that the shape of the entire universe might be dodecahedral. Although this would be amazing if true, subsequent research by other scientists has so far failed to turn-up evidence in support of this hypothesis.

Comparison of the 5 Platonic Solids

The details of each Platonic solid, including the total number of vertices, the number of faces, the number of edges, the number of edges on each face (the number of sides on each polygonal face), and the number of faces meeting at each vertex, is summarized in the table below:

Name	Vertices	Faces	Edges	Edges per Face	Faces meeting at vertex
Tetrahedron	4	4	6	3	3
Octahedron	6	8	12	3	4
Icosahedron	12	20	30	3	5
Hexahedron	8	6	12	4	3
Dodecahedron	20	12	30	5	3

Mathematicians formalize the information in this table, so that they can do further work with it (some of which we will see later).

Firstly, they assign the symbols V, F, and E, to representative the number of vertices, faces, or edges in a particular polyhedron.

Secondly, they make the following observations:

- **All** faces of Platonic solids are convex regular polygons.

- Faces of Platonic solids **only** intersect at their edges.

- In Platonic solids, the same number of faces must meet at **all** vertices.

A Platonic solid can thus be denoted by the following information in combination:

- The number of edges surrounding each face (which is also the number of vertices on each face), which is assigned the symbol p.

- The number of faces meeting at each vertex (which is also the number of edges intersecting at each vertex), which is assigned the symbol q.

Thus the combination of p and q together can be used to denote a particular Platonic solid. Mathematicians write this combination as $\{p, q\}$, and they call it a Schläfli symbol.

An alternative notation that mathematicians is also use to is known as **vertex configuration**. This simply lists the sidedness of the polygons around each vertex. So for example 3.3.3.3 means four 3-sided (triangular) faces meet at each vertex.

Here is a table of these mathematical attributes for the Platonic solids:

Name	V	F	E	Schläfli symbol	Vertex configuration
Tetrahedron	4	4	6	{3,3}	3.3.3
Octahedron	6	8	12	{3,4}	3.3.3.3
Icosahedron	12	20	30	{3,5}	3.3.3.3.3
Hexahedron	8	6	12	{4,3}	4.4.4
Dodecahedron	20	12	30	{5,3}	5.5.5

Only Five Types of Platonic Solid: Euclid's Proof

The first proof that there are only 5 possible Platonic solids dates back to the ancient Greeks, and may have been created by Theaetetus of Athens (c. 417 BC to 369 BC). None of Theaetetus' own writings have survived, but we know of him from one of Plato's dialogues about the nature of knowledge.

The earliest **surviving** proof that there are only 5 types of Platonic solids comes from Euclid of Alexandria, a Greek mathematician who lived in Egypt and is sometimes known as the "Father of Geometry" because of his great contributions to this field. This proof appears in Euclid's treatise entitled *Elements*, which was written in about 300 BC. We don't know whether the original source of proof that appears in *Elements*, but it was most likely based on Theaetetus' work.

A statue of Euclid in the Oxford University Museum of Natural History, Oxford, England:

The proof given in *Elements* goes something like this:

(1) Each face in a regular polyhedron must be a regular polygon.

(2) Each vertex in a polyhedron is formed at a junction of 3 or more faces.

(3) Each polygonal face must contribute an equal number of degrees to the junction at a vertex of a polyhedron. The number of degrees that each face contributes to the junction is given by the angle of the vertices within the polygon.

(4) Because we want a convex shape, the total number of degrees in the vertices of the polygons meeting at a junction must be **less** than 360°. Note: It can **not** be exactly 360° since this would result in a planar surface (flat surface).

(5) An equilateral triangle (3-sided regular polygon) has a 60° angles at its vertices, and so the possible combinations around a polyhedron's vertex are 3 equilateral triangles (resulting in a tetrahedron), 4 equilateral triangles (resulting in an octahedron), or 5 equilateral triangles (resulting in an icosahedron).   It is **not** possible to have a junction of 6 equilateral triangles since this would result in a planar surface, since 6 × 60° = 360°. Likewise, it is **not** possible to have a junction of 7 or more equilateral triangles, because this would exceed 360°.

(6) A square (4-sided regular polygon) has 90° angles at its vertices, and so the only possible combination around a junction is 3 squares (resulting in a regular hexahedron). It is **not** possible to have a junction of 4 squares since this would result in a planar surface, since 4 × 90° = 360°. Likewise, it is **not** possible to have a junction of 5 or more squares, because this would exceed 360°.

(7) A regular pentagon (5-sided regular polygon) has 108° angles at its vertices, and so the only possible combination of regular pentagons around a junction is 3 (resulting in a dodecahedron). It is **not** possible to have a junction of 4 or more regular pentagons, because would exceed 360°.

(8) A regular hexagon (6-sided regular polygon) has 120° angles at its vertices, and polygons with 7 or more sides have even larger angles. A junction of 3 or more of these polygons will always be at least 360° degrees, so it is **not** possible to form convex regular polyhedra using any of these shapes.

(9) We can thus see there are only five suitable shapes:

- A convex regular polyhedron with vertices formed by the junction of 3 equilateral triangles (this is of course the tetrahedron)

- A convex regular polyhedron with vertices formed by the junction of 4 equilateral triangles (this is of course the octahedron)

- A convex regular polyhedron with vertices formed by the junction of 5 equilateral triangles (this is of course the icosahedron)

- A convex regular polyhedron with vertices formed by the junction of 3 squares (this is of course the regular hexahedron or cube)

- A convex regular polyhedron with vertices formed by the junction of 3 regular pentagons (this is of course the dodecahedron)

Only Five Types of Platonic Solid: Topological Proof

Topology is a mathematical study of shapes and spaces, and is concerned with the properties of shapes/spaces which are preserved when the shape is bent or stretched, but **not** torn, cut, or glued (these changes of changes to the shape/space are known as "continuous deformations").

Topology gradually developed from geometry and set theory starting from about the 17th century, although the term "topology" was only coined in mid 19th century by Johann Benedict Listing (July 25th, 1808 to December 24th, 1882). Since at least the mid 20th century, topology has developed into a major branch of mathematics.

There is a topological proof that there are only 5 Platonic solids.

The key to the topological proof is the Euler characteristic of any convex polyhedron. The Euler characteristic of a shape/space, denoted by the Greek letter chi which has symbol χ, is a number which describes the shape/space regardless of how it might be bent or deformed by continuous deformations.

A portrait of Leonhard Euler (April 15th, 1707 to September 18th, 1783) appears on this Swiss 10 Franc note. Euler was one of the greatest mathematicians in history, and it is after him that the Euler characteristic is named.

The formula for χ is:

$$\chi = V - E + F$$

where *V* is the number of vertices in the shape/space, *E* the number of edges, and *F* the number of faces.

In the case of **any** convex polyhedron, the Euler characteristic will always be 2, so $\chi = 2$, or:

$$\chi = V - E + F = 2$$

You can see why this must be true by examining any existing convex polyhedron, such as a cube:

- If you count the number of vertices, edges, and faces, you will see that $\chi = 2$:

- If you imagine what would be the effect of adding an extra edge (for example the yellow diagonal line), you can see that you would also have to create a new face by splitting an existing face into two (labeled A and B in the picture), thus preserving $\chi = 2$:

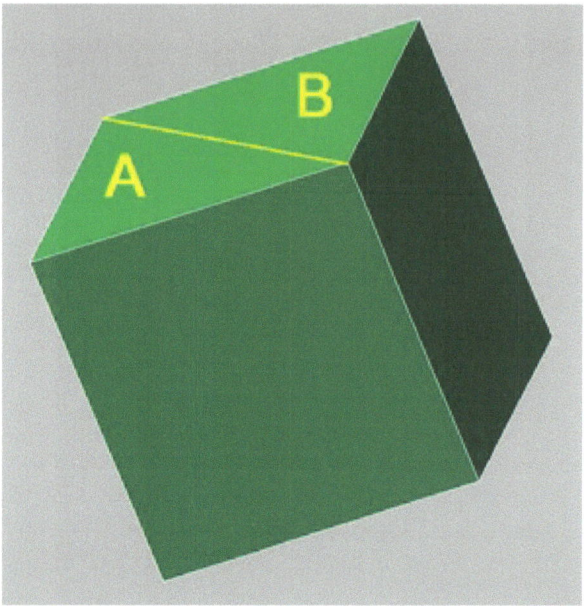

- Conversely if you were somehow to remove an edge, you would also merge two faces into one, thus reducing the number of faces by one, and again preserving $\chi = 2$.

- If you imagine what would be the effect of adding an extra vertex (for example at the orange cross), you can see that you would also have to create a new edge by splitting an existing edge into two (labeled C and D in the picture), thus preserving $\chi = 2$:

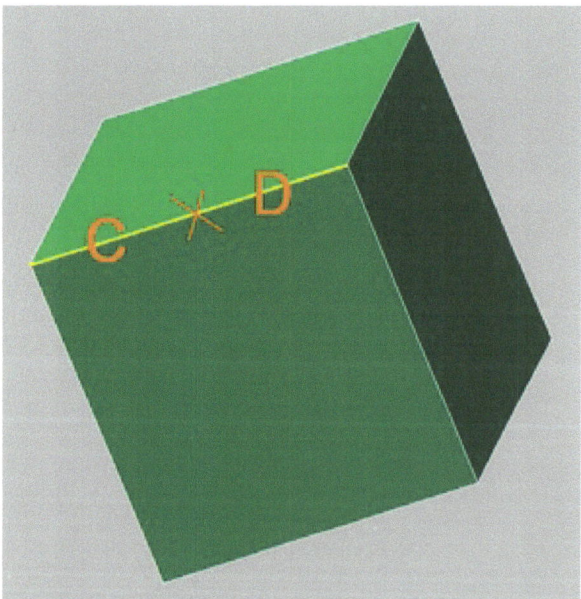

- Conversely if you were somehow to remove a vertex, you would also merge two edges into one, thus reducing the number of edges by one, and again preserving $\chi = 2$.

We can also observe another relationship between vertices, edges, and faces – expressed in two equations:

$$F = \frac{2E}{p}$$

$$V = \frac{2E}{q}$$

where again V is the number of vertices in the shape/space, E the number of edges, and F the number of faces – additionally, p is the number of edges on each face, and q is the number of edges meeting at each vertex.

A little bit of thought shows why both these equations must be true:

- You need p edges to border each face. So you might initially assume that the number of faces (F) would be given by ($E \div p$). However, each edge actually serves as the edge of two faces at the same time, hence you can double the number giving ($2E \div p$).

- Likewise, you need q edges to join together at each vertex. So you might initially assume that the number of vertices (V) would be given by ($E \div q$). However, each edge actually has two ends, and thus contributes to two vertices, so you can double the number giving ($2E \div q$).

We can now substitute the formulas for F and V into the earlier equation for χ, giving:

$$\chi = \frac{2E}{q} - E + \frac{2E}{p} = 2$$

Or just:

$$\frac{2E}{q} - E + \frac{2E}{p} = 2$$

Rearranging this equation leads us to:

$$\frac{1}{q} + \frac{1}{p} = \frac{1}{2} + \frac{1}{E}$$

Now, since we know that E is always positive, this means that $\dfrac{1}{E}$ is also always positive. We can therefore reach this inequality:

$$\frac{1}{q} + \frac{1}{p} > \frac{1}{2}$$

Additionally, we know that p and q must both be positive integers (whole numbers) greater than or equal to 3. By a simple process of trying out different values of p and q, we can quickly see there are only 5 possibilities for (p, q) – namely (3, 3), (3, 4), (3, 5), (4, 3), (5, 3), which of course correspond to the 5 Platonic solids, and their Schläfli symbols.

I have reproduced the table of Platonic solids with their Schläfli symbols here:

Name	V	F	E	Schläfli symbol	Vertex configuration
Tetrahedron	4	4	6	{3,3}	3.3.3
Octahedron	6	8	12	{3,4}	3.3.3.3
Icosahedron	12	20	30	{3,5}	3.3.3.3.3
Hexahedron	8	6	12	{4,3}	4.4.4
Dodecahedron	20	12	30	{5,3}	5.5.5

Dual Polyhedra

Every type of polyhedron has an associated dual polyhedron (also known as a "polar polyhedron") which is made by putting the center of face where each vertex was, and a vertex where the center of each face was – thus exchanging vertices and faces.

In the case of Platonic solids, each polyhedron's dual is always a Platonic solid.

- The dual polyhedron of a regular tetrahedron is another regular tetrahedron, albeit rotated by 180°. Mathematicians describe this by saying a regular tetrahedron is self-dual.

- The dual of a regular octahedron is a regular hexahedron (cube), and vice-versa.

- The dual of a regular icosahedron is a regular dodecahedron, and vice-versa.

A regular hexahedron (cube) is the dual polyhedron of a regular octahedron.

Note that each vertex of the octahedron corresponds to a center of a face in the hexahedron.

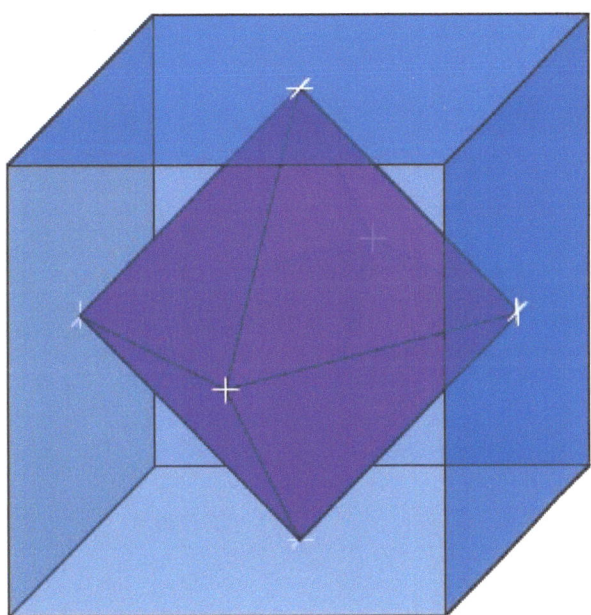

One thing that you might notice is that because a dual polyhedron has faces and vertices exchanged, if {p, q} is the Schläfli symbol of the starting polyhedron, then {q, p} will be the Schläfli symbol of its dual. This is shown in the table below, with each Platonic solid listed on the left (blue background), and its dual on the right (orange background):

Name	Schläfli symbol	Dual	Schläfli symbol
Tetrahedron	{3,3}	Tetrahedron	{3,3}
Octahedron	{3,4}	Hexahedron	{4,3}
Icosahedron	{3,5}	Dodecahedron	{5,3}
Hexahedron	{4,3}	Octahedron	{3,4}
Dodecahedron	{5,3}	Icosahedron	{3,5}

More generally, we can see the numbers associated with several different properties are exchanged when you compare each Platonic solid's attributes to those of its dual polyhedron. As already noted, faces and vertices are swapped, but we can also see swaps between the number of faces meeting at each vertex, and the number of edges around each face.

These relationships between each polyhedron and its respective dual are shown in the following tables:

	Hexahedron	Octahedron
Faces	6	8
Vertices	8	6
Edges	12	12
Edges per Face	4	3
Faces meeting at vertex	3	4

	Dodecahedron	Icosahedron
Faces	12	20
Vertices	20	12
Edges	30	30
Edges per Face	5	3
Faces meeting at vertex	3	5

	Tetrahedron	Tetrahedron
Faces	4	4
Vertices	4	4
Edges	6	6
Edges per Face	3	3
Faces meeting at vertex	3	3

Calculating Angles in Platonic Solids

In this chapter I will talk about some of the angular properties of Platonic solids (I have previously simply given figures for some of the angles without explanation). The mathematics involved is slightly more complex that in most other parts of this book, so before doing so I will **briefly** explain some necessary mathematical terminology and concepts – however should you wish to study these concepts in more depth, I would suggest that you do some further reading on trigonometry.

You should also be aware of the number π (written in English characters as "pi" and pronounced like the type of pie that you might eat) which appears in many formulae in this book. π is a famous mathematical constant with a value of approximately 3.14159265. For more information on π, see my book **Amazing Math:** The Most Interesting, Astonishing, and Absolutely Awesome Numbers in the Universe.

Radians & Steradians

Note: I will present a brief discussion of radians and steradians here, but you may also wish to see my book **You Can Do Math: Radians and Steradians**.

Most people are familiar with measuring angles in degrees. 360° is a full-circle rotation, a half-circle rotation is 360° ÷ 2 = 180°, a quarter-turn 360° ÷ 4 = 90°, and so on.

The choice of 360 as representing a full-circle rotation is arbitrary – this number was chosen by the Babylonians as it is easy to work with. However, we could equally well choose a different number – for example there is a French system called gradians which uses 400 to represent a full-circle rotation.

Mathematicians and scientists do use degrees, but they also often use a different unit to measure angles: radians. In radians, a full-circle rotation is 2π radians. The reason that they chose this number is it makes many mathematical formulae simpler and easier to work with.

For example, if we imagine an arc that covers part of the circumference of a circle – the angle spanned by the arc (labeled α in the diagram) in radians, is the ratio between the length of the arc (labeled b in the diagram) and the radius of the circle (labeled r in the diagram). Or, to put it another way, if we are measure angles in radians, $\alpha = b \div r$:

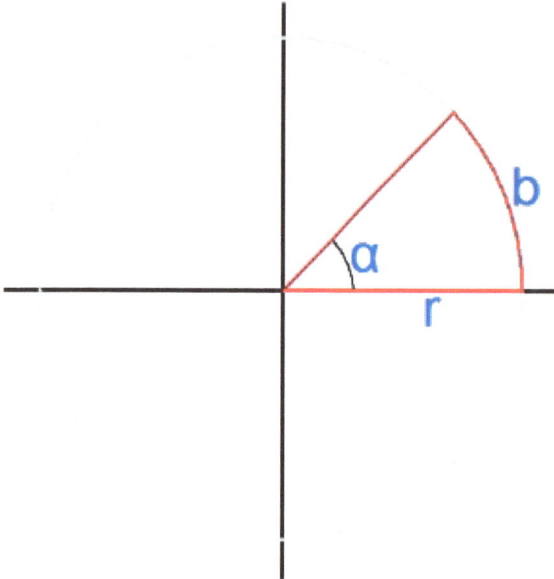

We thus define radians: the angle subtended at the centre of a circle measured in radians is equal to length of the arc divided by the radius of the circle.

Hopefully you can see that since 360° is the same as 2π radians, 180° must be the same as π radians, 90° the same as $\pi \div 2$ radians, and so on.

Measuring angles in radians (or degrees or even gradians) is fine in two dimensions, but in three dimensions we need a different unit. The unit mathematicians use in this case is steradians.   Just as a radian is the ratio of an arc at the circumference of a circle and the circle's radius, a steradian is the ratio of a c one at the surface of a sphere and the sphere's radius.

Trigonometric Functions

Trigonometric functions are functions based on the ratio of the lengths of sides in a right triangle (British English: right-angled triangle).

The sides are in the triangle are labeled:

- The hypotenuse is the longest side, and is always opposite the right angle.

- The adjacent is the side connecting the chosen angle (marked Θ in the diagram below) to the right angle.

- The opposite is the side which is opposite the chosen angle (marked Θ in the diagram below).

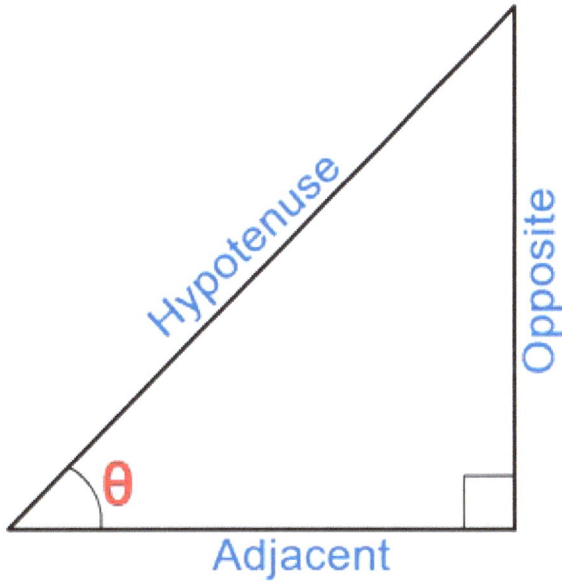

The trigonometric functions are then defined:

- **sine** (often abbreviated to **sin**) is the ratio of the Opposite to the Hypotenuse. That is to say sin Θ = Opposite ÷ Hypotenuse

- **cosine** (often abbreviated to **cos**) is the ratio of the Adjacent to the Hypotenuse. That is to say cos Θ = Adjacent ÷ Hypotenuse

- **tangent** (often abbreviated to **tan**) is the ratio of the Opposite to the Adjacent. That is to say tan Θ = Opposite ÷ Adjacent

- **cosecant** (often abbreviated to **cosec**) is the ratio of the Hypotenuse to the Opposite. That is to say cosec Θ = Hypotenuse ÷ Opposite

- **secant** (often abbreviated to **sec**) is the ratio of the Hypotenuse to the Adjacent. That is to say sec Θ = Hypotenuse ÷ Adjacent

- **cotangent** (often abbreviated to **cot**) is the ratio of the Adjacent to the Opposite. That is to say cot Θ = Adjacent ÷ Opposite

As mentioned previously, if you wish to study these concepts in more depth, I would suggest you do some further reading on **trigonometry**.

Applying Trigonometry to Platonic Solids

There are various mathematical formulae which describe the angles in Platonic solids, and each of which is connected to the Schläfli symbol, {*p, q*}, of the solid in question.

The **face angle**, which I have denoted using the symbol α, is the angle at each vertex on each polygonal face:

$$\alpha \; = \; \pi \; - \; (\; 2 \; \pi \; \div \; p \;)$$

The **dihedral angle**, usually denoted by the symbol Θ, is the interior angle between any two faces. It can be calculated using this formula:

$$\sin \frac{\Theta}{2} \; = \; \frac{\cos(\pi \div q)}{\sin(\pi \div p)}$$

The dihedral angle can also be calculated using this formula:

$$\tan \frac{\Theta}{2} \; = \; \frac{\cos(\pi \div q)}{\sin(\pi \div h)}$$

where *h* (known as the **Coxeter number**) is 4 for a tetrahedron, 6 for a hexahedron or octahedron, and 10 for a tetrahedron or icosahedron.

The **angular deficiency**, usually denoted by the symbol δ, is the difference between the sum of the face angles and 2π.

$$\delta \; = \; 2 \; \pi \; - \; q \; \pi \; (\; 1 \; - \; \frac{2}{p} \;)$$

However there is also a simpler formula for δ. French mathematician and philosopher René Descartes (March 31st, 1596 to February 11th, 1650) showed that the total angular deficiency will always be 4π, hence:

$$\delta \; = \; 4 \; \pi \; \div \; V$$

where *V* is the total number of vertices in the Platonic solid.

Finally, the **solid angle** at each vertex, usually denoted by the symbol Ω, can be calculated from the dihedral angle (Θ) using this formula:

$$\Omega \; = \; q \; \Theta \; - \; \pi \; (\; q \; - \; 2 \;)$$

Platonic Solids and Spheres

Because they are regular, all Platonic solids are linked to three related spheres:

- The **circumscribed sphere** is a sphere that passes through all the solid's vertices. The radius of the circumscribed sphere is referred to as the **circumradius**, and is usually denoted using the symbol R.

- The **midsphere** is a sphere that touches ("is tangent to") each of the solid's edges at the midpoint of the edge. The radius of the midsphere is referred to as the **midradius**, and is usually denoted by the Greek letter rho which has symbol ρ.

- The **inscribed sphere** is a sphere that touches ("is tangent to") each face at the center of the face. The radius of the inscribed sphere is referred to as the **inradius**, and is usually denoted using the symbol r.

The following formulae can be used for calculating each of these radii:

$$R = \left(\frac{a}{2}\right) \tan\frac{\pi}{q} \tan\frac{\Theta}{2}$$

where Θ is the dihedral angle (see Calculating Angles in Platonic Solids).

$$\rho = \left(\frac{a}{2}\right) \frac{\cos(\pi \div q)}{\sin(\pi \div h)}$$

where h (known as the **Coxeter number**) is 4 for a tetrahedron, 6 for a hexahedron or octahedron, and 10 for a tetrahedron or icosahedron.

$$r = \left(\frac{a}{2}\right) \cot\frac{\pi}{p} \tan\frac{\Theta}{2}$$

where again, Θ is the dihedral angle.

Platonic Solid Surface Area

The following formula can be used for calculating the surface area of a Platonic solid:

$$A = \left(\frac{a}{2} \right)^2 F p \cot \frac{\pi}{p}$$

where F is the number of faces.

Platonic Solids Volume

The following formula can be used for calculating the volume of a Platonic solid:

$$V = \frac{1}{3} \, r \, A$$

Other Interesting Types of Polyhedra

In this book, we have talked about Platonic solids which are convex regular polyhedra. There are many other polyhedra as well.

Here are some of the most interesting other types of polyhedra:

(1) There are **prisms**, which consist of two parallel congruent faces (known as **"bases"**), both in the same rotational orientation, and a series of other faces (which must necessarily all be parallelograms or rectangles) joining the corresponding sides of the two bases.

Here is an example of a prism:

(2) There are **antiprisms**, which consist of two parallel congruent faces (again called **"bases"**), which are twisted relative to each other, and which are connected by an alternating band of triangles.

Here is an example of an antiprism:

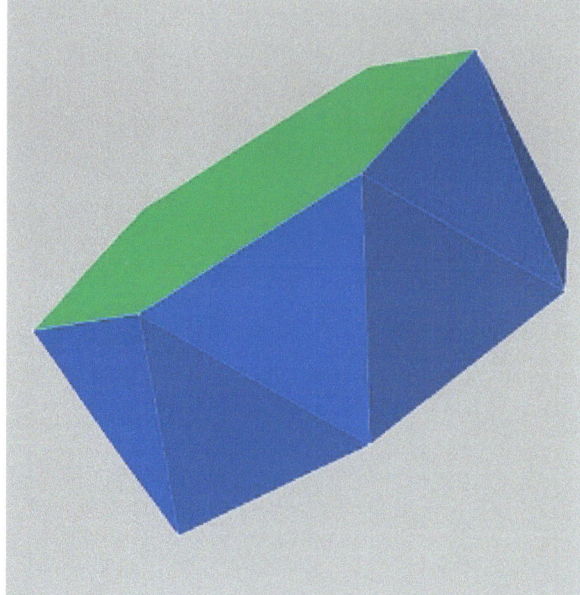

 (3) There are 4 types of polyhedra which are regular (like Platonic solids), but which are **not** convex (unlike Platonic solids). These are known as the "[Kepler-Poinsot polyhedra](#)". Each of these polyhedra has a star-like shape, so they are sometimes called "star polyhedra".

Two of these star polyhedra were discovered by the Middle Ages or possibly earlier, although it is not clear who discovered them. The other two star polyhedra were discovered by French mathematician Louis Poinsot (January 3rd, 1777 to December 5th, 1859) in 1809.

After his discovery, Louis Poinsot was unsure if there were any more types of star polyhedra. However, just three years after Poinsot's discovery, Baron Augustin-Louis Cauchy (August 21st, 1789 to May 23rd, 1857) proved there were only four types.

The star polyhedra are today collectively named after Johannes Kepler (December 27th, 1571 to November 15th, 1630) who was first to recognize the two then-known polyhedra as regular, and Louis Poinsot, hence the name Kepler-Poinsot polyhedra.

Here are the Kepler-Poinsot polyhedra:

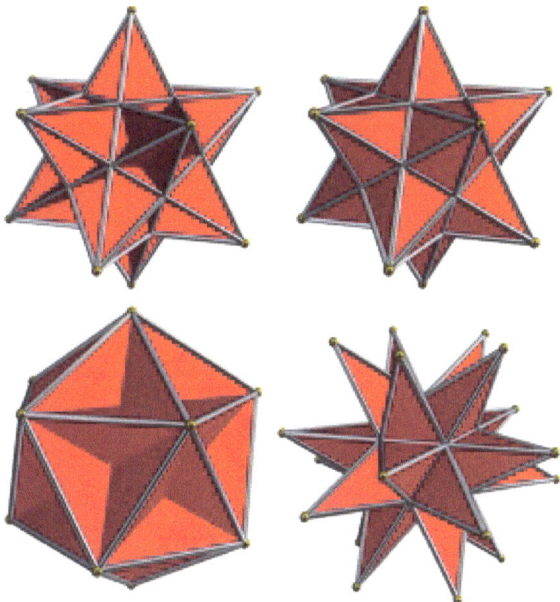

Here are the names of the Kepler-Poinsot polyhedra (in the same order as the illustration above, starting from the top-left, and then horizontally across each row, then vertically):

- Great icosahedron

- Small stellated dodecahedron

- Great dodecahedron

- Great stellated dodecahedron

(4) There are 13 types of **Archimedean solids** (note: 2 of the 13 types are "**chiral**" or "**enantiomorphic**" – they have mirror images which differ from the original – so if we were to count the enantiomorphs separately, we could say there are 15 types). Archimedean solids are convex polyhedra which are composed of two or more types of regular polygons, always meeting in identical vertices.

The Archimedean solids are named after the Greek philosopher, engineer and mathematician Archimedes (c. 287 BC to c. 212 BC) who discussed them in a work that has been subsequently lost. We know that Archimedes was aware of all 13 types, because Pappus of Alexandria (c. 290 to c. 350 AD) mentions this in his writings.

Here are the Archimedean solids:

Here are the names of the Archimedean solids (in the same order as the illustration above, starting from the top-left, and then horizontally across each row, then vertically):

- Truncated octahedron

- Truncated cube

- Cuboctahedron

- Truncated tetrahedron

- Icosidodecahedron

- Snub cuboctahedron

- Rhombitruncated cuboctahedron

- Rhombicuboctahedron

- Rhombitruncated Icosidodecahedron

- Rhombicosidodecahedron

- Truncated icosahedron

- Truncated dodecahedron

- Snub Icosidodecahedron

(5) There are 13 types of <u>Catalan solids</u> (or 15 types if we count enantiomorphs separately, since 2 of the 13 types are chiral). The Catalan solids are the dual polyhedra for the Archimedean solids, and have faces which are **not** regular polygons.

The Catalan solids are named after the Belgian mathematician Eugène Catalan (May 30th, 1814 to February 14th, 1894) who first described them in 1865.

Here are the Catalan solids:

Here are the names of the Catalan solids (in the same order as the illustration above, starting from the top-left, and then horizontally across each row, then vertically):

- Triakis hexahedron

- Triakis octahedron

- Rhombic dodecahedron – Note: you may recall from earlier in this book, that when people say that a diamond or garnet exhibits "dodecahedral habit", they are referring to this shape.

- Triakis tetrahedron

- Rhombic triacontahedron

- Pentagonal icositetrahedron

- Disdyakis dodecahedron

- Deltoidal icositetrahedron

- Disdyakis triacontahedron

- Deltoidal hexecontahedron

- Pentakis dodecahedron

- Triakis icosahedron

- Pentagonal hexecontahedron

Garnet crystal with dodecahedral habit (rhombic dodecahedron shaped):

(6) There are Johnson solids. There are convex polyhedra where the faces are two or more types of regular polygon, joined together in any way, and which are not Platonic solids, nor Archimedean solids, nor prisms, nor antiprisms.

These solids are named after Norman W. Johnson, an American mathematician, who in 1966 enumerated a list of 92 solids, and conjectured (but did **not** prove) that his list was complete. Three years later in 1969, a Russian mathematician, Viktor Abramovich Zalgaller, proved that Johnson's list was complete.

As all Johnson solids' faces are regular polygons, the smallest angle in any regular polygon is 60° (in an equilateral triangle), and the sum of the angles in the joining faces at a vertex must be less than 360° (since Johnson solids are convex), there is a maximum of five faces meeting at any vertex in a Johnson solid.

There is no obvious restriction on the number of sides each polygonal face of a Johnson solid must have, however it does turn out that the faces can only have 3, 4, 5, 6, 8 or 10 sides.

An example of a Johnson solid is a square pyramid – it has 1 face which is a square, and 4 faces which are equilateral triangles:

Since there are only 92 Johnson solids, each is given both a name and an identifying number. Here is the full list of Johnson solids:

- J1 – Square pyramid

- J2 – Pentagonal pyramid

- J3 – Triangular cupola

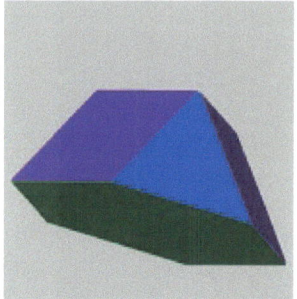

- J4 – Square cupola

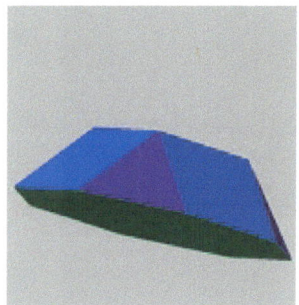

- J5 – Pentagonal cupola

- J6 – Pentagonal rotunda

- J7 – Elongated triangular pyramid

- J8 – Elongated square pyramid

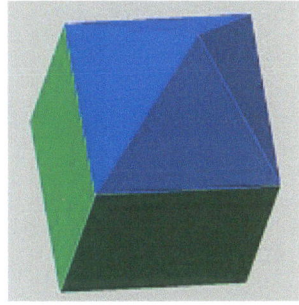

- J9 – Gyroelongated triangular pyramid

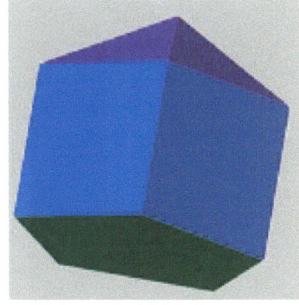

- J10 – Gyroelongated square pyramid

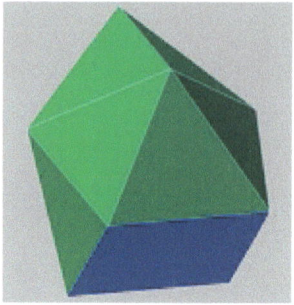

- J11 – Gyroelongated pentagonal pyramid

- J12 – Triangular bipyramid

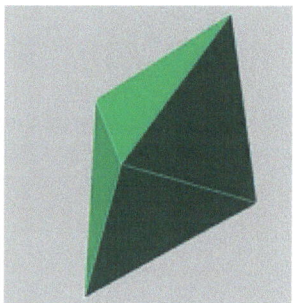

- J13 – Pentagonal bipyramid

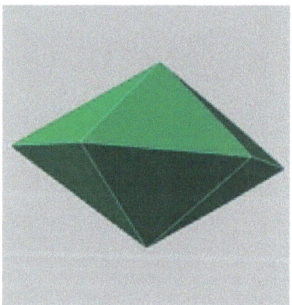

- J14 – Elongated triangular bipyramid

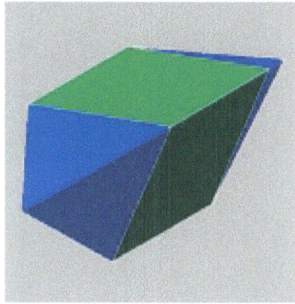

- J15 – Elongated square bipyramid

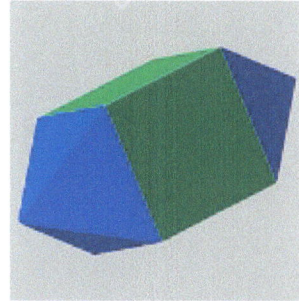

- J16 – Elongated pentagonal bipyramid

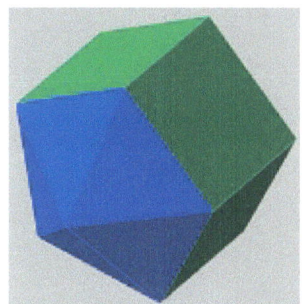

- J17 – Gyroelongated square bipyramid

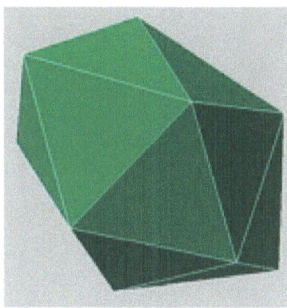

- J18 – Elongated triangular cupola

- J19 – Elongated square cupola

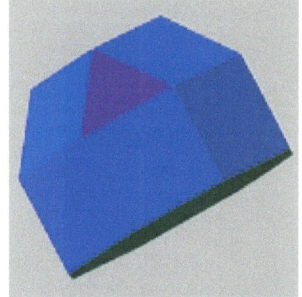

- J20 – Elongated pentagonal cupola

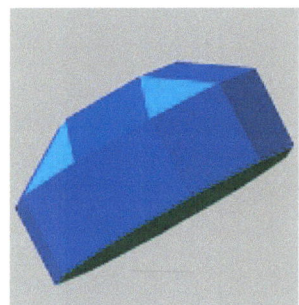

- J21 – Elongated pentagonal rotunda

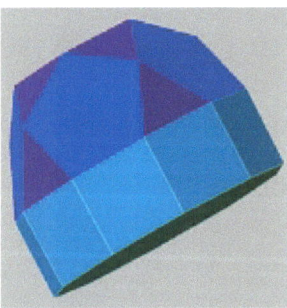

- J22 – Gyroelongated digonal cupola

- J23 – Gyroelongated square cupola

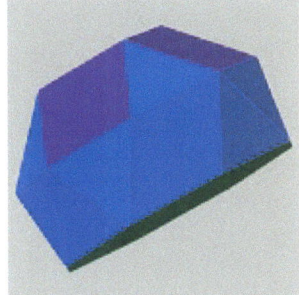

- J24 – Gyroelongated pentagonal cupola

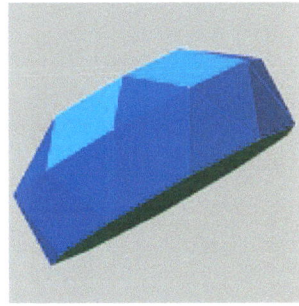

- J25 – Gyroelongated pentagonal rotunda

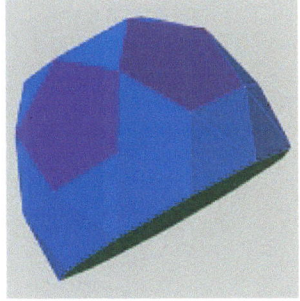

- J26 – Digonal gyrobicupola (gyrobifastigium)

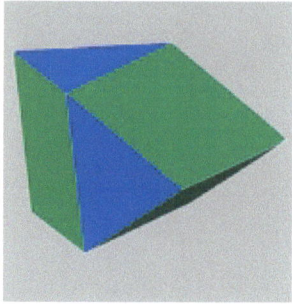

- J27 – Triangular orthobicupola

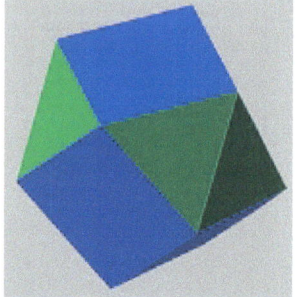

- J28 –Square orthobicupola

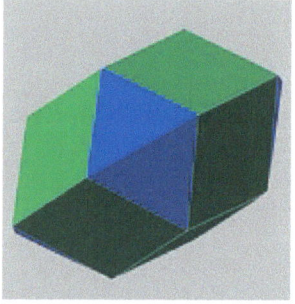

- J29 – Square gyrobicupola

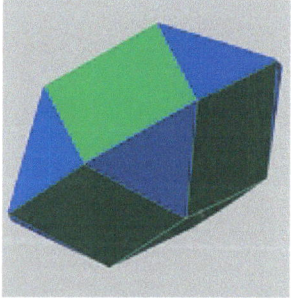

- J30 – Pentagonal orthobicupola

- J31 – Pentagonal gyrobicupola

- J32 – Pentagonal orthocupolarotunda

- J33 – Pentagonal gyrocupolarotunda

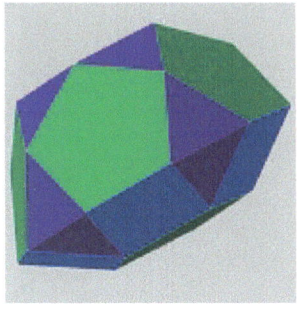

- J34 – Pentagonal orthobirotunda

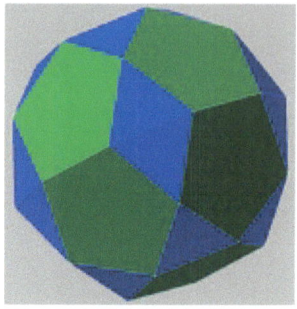

- J35 – Elongated triangular orthobicupola

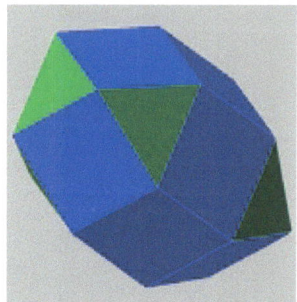

- J36 – Elongated triangular gyrobicupola

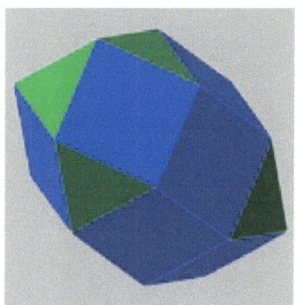

- J37 – Elongated square gyrobicupola

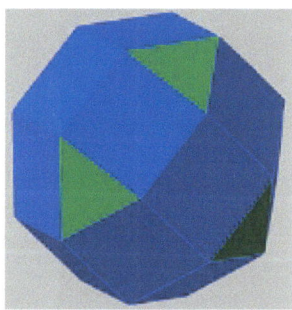

- J38 – Elongated pentagonal orthobicupola

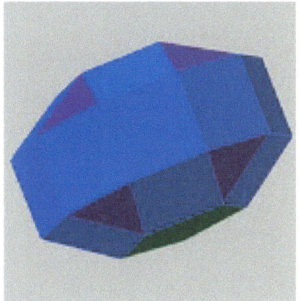

- J39 – Elongated pentagonal gyrobicupola

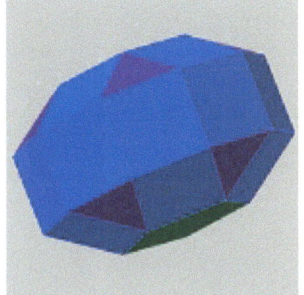

- J40 – Elongated pentagonal orthocupolarotunda

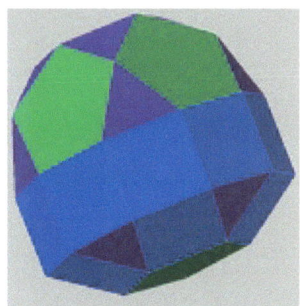

- J41 – Elongated pentagonal gyrocupolarotunda

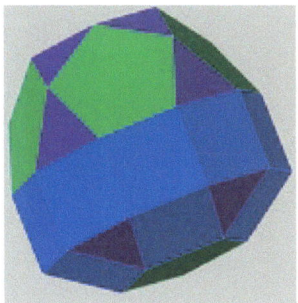

- J42 – Elongated pentagonal orthobirotunda

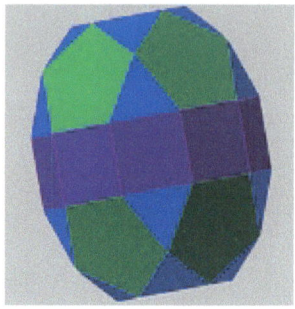

- J43 – Elongated pentagonal gyrobirotunda

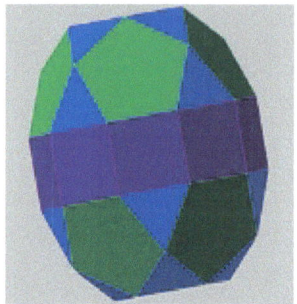

- J44 – Gyroelongated triangular bicupola

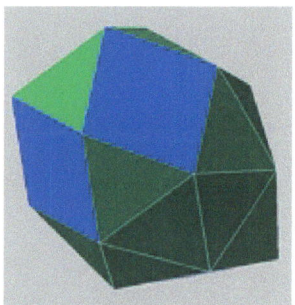

- J45 – Gyroelongated square bicupola

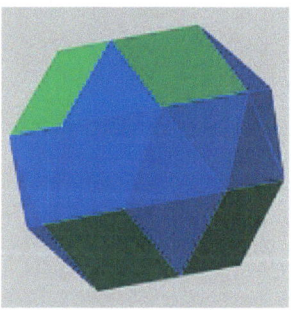

- J46 – Gyroelongated pentagonal bicupola

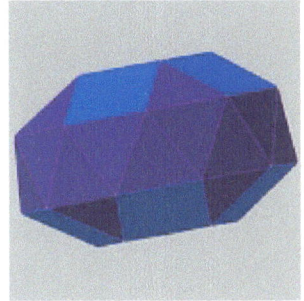

- J47 – Gyroelongated pentagonal cupolarotunda

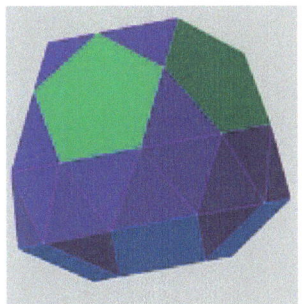

- J48 – Gyroelongated pentagonal birotunda

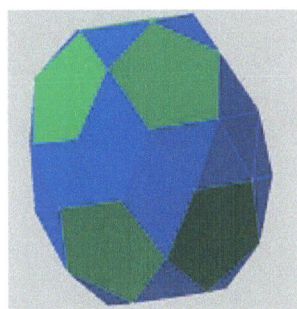

- J49 – Augmented triangular prism

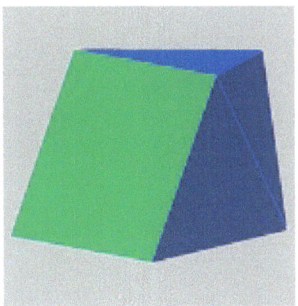

- J50 – Biaugmented triangular prism

- J51 – Triaugmented triangular prism

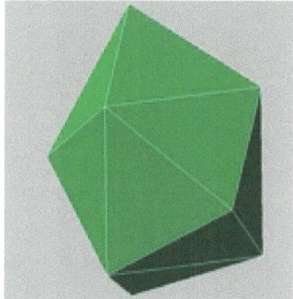

- J52 – Augmented pentagonal prism

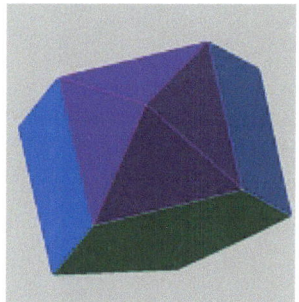

- J53 – Biaugmented pentagonal prism

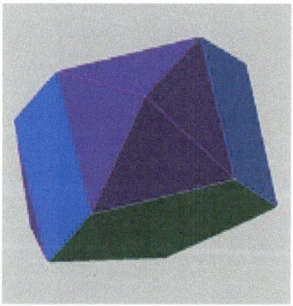

- J54 – Augmented hexagonal prism

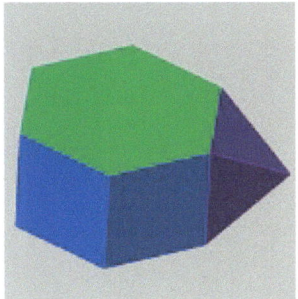

- J55 – Parabiaugmented hexagonal prism

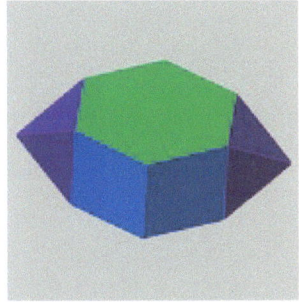

- J56 – Metabiaugmented hexagonal prism

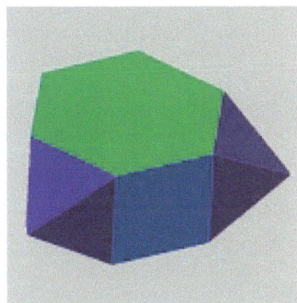

- J57 – Triaugmented hexagonal prism

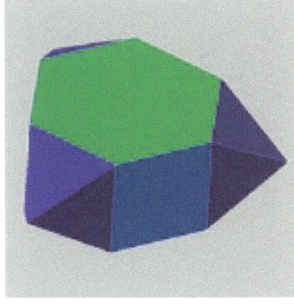

- J58 – Augmented dodecahedron

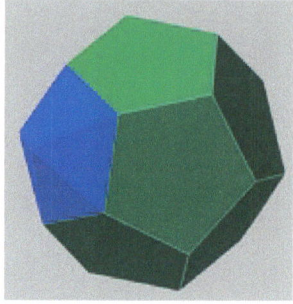

- J59 – Parabiaugmented dodecahedron

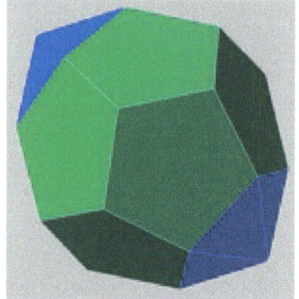

- J60 – Metabiaugmented dodecahedron

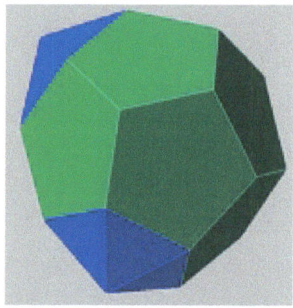

- J61 – Triaugmented dodecahedron

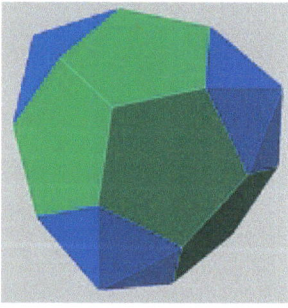

- J62 – Metabidiminished icosahedron

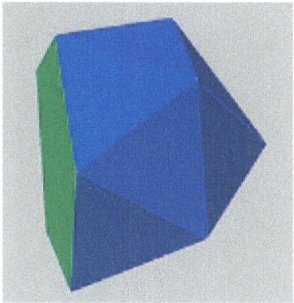

- J63 – Tridiminished icosahedron

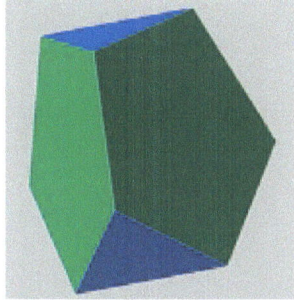

- J64 – Augmented tridiminished icosahedron

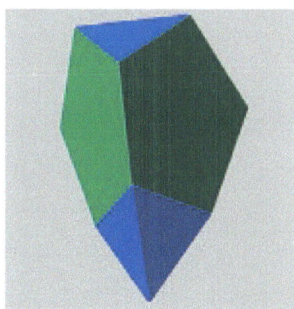

- J65 – Augmented truncated tetrahedron

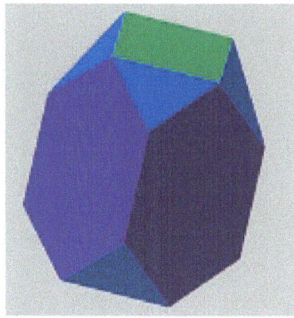

- J66 – Augmented truncated cube

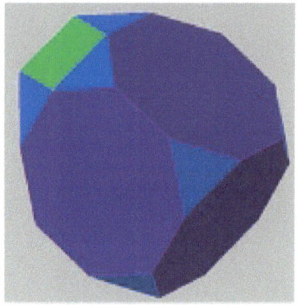

- J67 – Biaugmented truncated cube

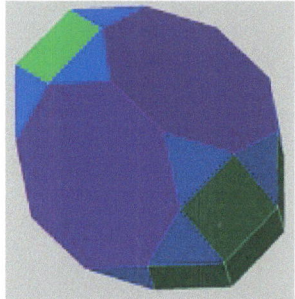

- J68 – Augmented truncated dodecahedron

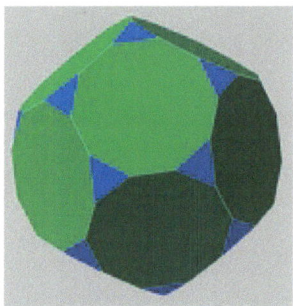

- J69 – Parabiaugmented truncated dodecahedron

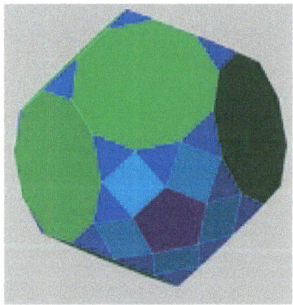

- J70 – Metabiaugmented truncated dodecahedron

- J71 – Triaugmented truncated dodecahedron

- J72 – Gyrate rhombicosidodecahedron

- J73 – Paragyrate rhombicosidodecahedron

- J74 – Metabigyrate rhombicosidodecahedron

- J75 – Trigyrate rhombicosidodecahedron

- J76 – Diminished rhombicosidodecahedron

- J77 – Paragyrate diminished rhombicosidodecahedron

- J78 – Metagyrate diminished rhombicosidodecahedron

- J79 – Bigyrate diminished rhombicosidodecahedron

- J80 – Paradiminished rhombicosidodecahedron

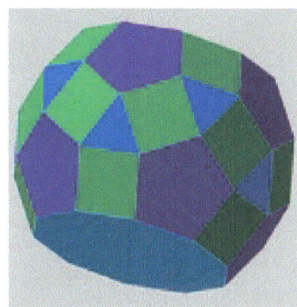

- J81 – Metabidiminished rhombicosidodecahedron

- J82 – Gyrate bidiminished rhombicosidodecahedron

- J83 – Tridiminished rhombicosidodecahedron

- J84 – Snub disphenoid

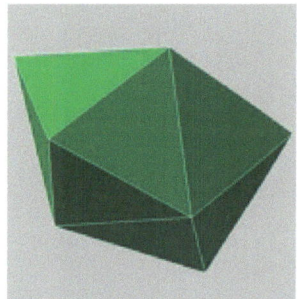

- J85 – Snub square antiprism

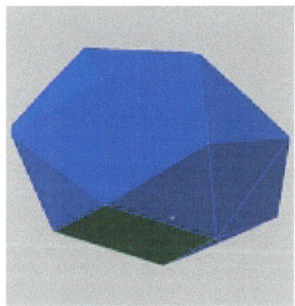

- J86 – Sphenocorona

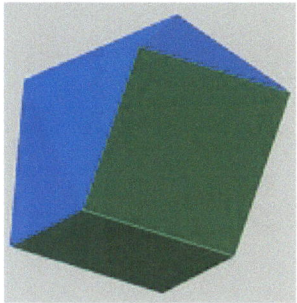

- J87 – Augmented sphenocorona

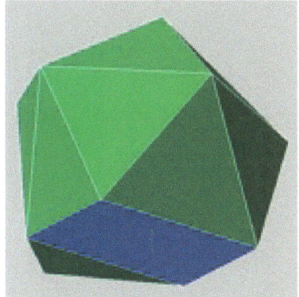

- J88 – Sphenomegacorona

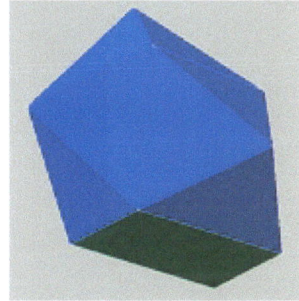

- J89 – Hebesphenomegacorona

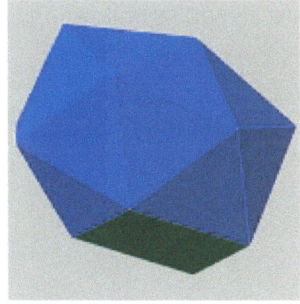

- J90 – Disphenocingulum

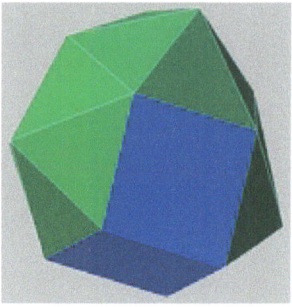

- J91 – Bilunabirotunda

- J92 – Triangular hebesphenorotunda

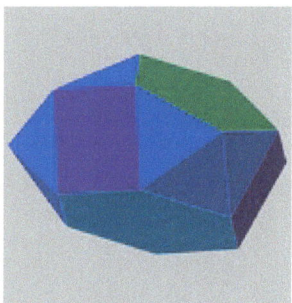

(7) Deltahedra (singular: deltahedron) are polyhedra whose faces are all equilateral triangles. There is an infinitely many deltahedra, but only 8 types of deltahedra which are convex:

- 3 of the 8 convex types are regular (Platonic solids): tetrahedron, octahedron, icosahedron – as shown in the illustration:

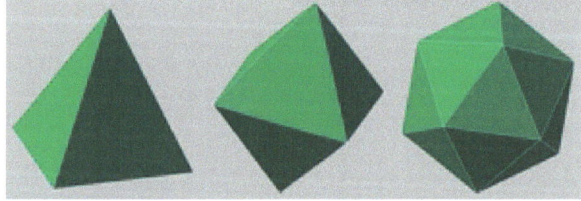

- 5 of the 8 convex types are Johnson solids: triangular bipyramid (J12), pentagonal bipyramid (J13), gyroelongated square bipyramid (J17), triaugmented triangular prism (J51), snub disphenoid (J84) – as shown in the illustration:

Conclusion

Well done for getting to the end. I hope you enjoyed this book!

For more Platonic solids (and to learn about other polyhedra too), please go to:

http://www.suniltanna.com/platonic

If you enjoyed this book or it helped you, please post a positive review on Amazon!

To find out about other educational books that I have written, please go to:

- For math books: http://www.suniltanna.com/math
- For science books: http://www.suniltanna.com/science

And remember: **If you enjoyed this book or it helped you, please post a positive review on Amazon!**

www.ingramcontent.com/pod-product-compliance
Lightning Source LLC
Chambersburg PA
CBHW050737180526
45159CB00003B/1263